Essential Artificial Intelligence for AI Rookies

Brief and easy to understand explanations.

Mikhail Van Starren
Author

Table of Contents

About this book

"Essential Artificial Intelligence for Rookies" is a unique and captivating journey into the fascinating world of AI, designed specifically for newcomers. This book unravels the complexities of AI, transforming them into engaging and digestible insights that make learning not just easy, but enjoyable. From the philosophical roots of AI to the innovative techniques powering today's most innovative technologies, we've distilled the essence of artificial intelligence into a guide that's comprehensive yet accessible. Whether you're planning to delve into a career in AI, or simply satiate your curiosity about the digital revolution transforming our world, this book is your ticket to a whole new understanding. Embark on this journey with us and see the world of AI illuminated like never before.

Dedication

"To our beloved baby, Kayden, who couldn't make it into this world. Even though we didn't have the chance to meet you, your brief presence in our lives has left a profound impact. Your mom and dad love you deeply and unconditionally. You will forever be in our hearts, cherished and remembered."

Introduction

Welcome to "Essential Artificial Intelligence for Rookies." If you've ever found yourself intrigued by Siri's responses, awed by autonomous vehicles, or marveling at the recommendations served by your favorite streaming service, then you're already familiar with the everyday applications of Artificial Intelligence (AI). AI, with its vast potential and rapidly advancing capabilities, has permeated every facet of our lives, revolutionized industries and redefining our experiences.

This book aims to demystify the complex world of AI, translating intricate concepts into digestible insights for rookies. Whether you are a student venturing into the realm of computer science, a professional aiming to stay abreast with technology's rapid pace, or simply a curious mind keen on understanding the digital revolution, this book is your passport to the exciting universe of AI.

We will guide you, step-by-step, through the fundamentals of AI, beginning with its roots and philosophical underpinnings, moving through the core concepts and techniques, and finally exploring its applications and ethical implications. We will delve into the realms of machine learning, deep learning, natural language processing, and computer vision, among others, always with an emphasis on understanding and accessibility.

Artificial Intelligence doesn't have to be an intimidating or inaccessible topic. With "Essential Artificial Intelligence for Rookies," you hold in your hands the key to unlock the secrets of this transformative technology. So, are you ready to step into the future? Let's embark on this exciting journey together.

What is Artificial Intelligence?

Artificial Intelligence (AI) refers to the simulation of human intelligence processes by machines, particularly computer systems. These processes include learning (acquiring information and the rules for using it), reasoning (using rules to reach approximate or definite conclusions), and self-correction.

John McCarthy, who is often credited as the father of AI, defined it as "the science and engineering of making intelligent machines." AI involves developing algorithms and methodologies that allow computers to perform tasks that would normally require human intelligence.

AI is classified into two types:

Narrow AI: These are systems designed to conduct specific tasks, such as voice recognition, recommendation systems, or image recognition. They operate under a limited context and can only do what they are programmed to do.

General AI: These systems can understand, learn, adapt, and implement knowledge in a broad range of tasks, just as a human would. As of my knowledge cut-off in 2021, this kind of AI, often referred to as "strong AI," is theoretical and doesn't yet exist.

Applications and Real-World Examples of AI

AI technologies are being used across a broad range of industries, impacting everyday life in many ways. Here are some notable applications:

Healthcare: AI is being used to predict disease, diagnose conditions, recommend treatments, and even assist in surgeries.

Automotive Industry: AI powers the technology behind autonomous vehicles. These vehicles rely on AI to understand their surroundings and make driving decisions.

E-commerce and Marketing: Companies like Amazon and Netflix use AI algorithms to offer personalized recommendations, enhancing customer experience and boosting sales.

Finance: Banks and other financial institutions use AI for fraud detection, risk assessment, investment prediction, customer service, and more.

Manufacturing: AI-powered robots are playing an essential role in automating production lines, resulting in increased productivity and accuracy.

Agriculture: AI is used to predict yield, monitor crop health, automate farming operations, and optimize resource usage.

The Future of AI and Ethical Considerations

The future of AI holds significant promise, and its impact is expected to be transformative. It's likely that AI will continue to be integrated into increased aspects of our lives. We can expect advancements in areas such as healthcare, where AI might be used for more accurate diagnoses and personalized treatment plans; transportation, where self-driving cars could become the norm; and education, where AI could customize the learning experience to individual needs.

However, with enormous potential comes great responsibility. AI also raises complex ethical and societal issues that need to be addressed. For instance, there are concerns about job displacement due to automation, security issues related to AI, privacy concerns arising from the collection and use of enormous amounts of data, and the potential for AI systems to perpetuate or even exacerbate societal biases.

A significant ethical concern involves the transparency and interpretability of AI models, also known as the "black box" problem. It refers to the lack of understanding of how complex AI models, like deep neural networks, make their decisions.

AI's potential is immense, but it must be developed and used responsibly, with an awareness of its impact on society. The conversation around AI ethics is as essential as the technology itself and must be ongoing to ensure a future in which AI serves the best interests of humanity.

Mathematics for Artificial Intelligence- The Foundation

In the world of Artificial Intelligence (AI), mathematics provides the bedrock upon which all AI systems are built. While the multidisciplinary field of AI draws from computer science, cognitive psychology, and linguistics, mathematics offers the necessary tools to model, analyze, and understand the core algorithms that give AI its intelligence. Here, we delve into the key mathematical concepts that underpin AI.

To fully grasp AI and its workings, one needs to have a fundamental understanding of the following areas of mathematics:

Linear Algebra: Provides the language to describe vector and matrix operations- the foundation for handling data and models in AI.

Calculus: Needed for optimization problems and understanding learning algorithms, especially in the context of Neural Networks.

Probability Theory and Statistics: Offers the framework for managing uncertainty and making informed decisions based on data.

Linear Algebra and AI

Linear Algebra deals with vectors, vector spaces, linear transformations, and systems of linear equations. It's a vital tool in the design of AI algorithms, mainly because of its efficiency in handling multi-dimensional data.

A significant part of AI, especially machine learning and deep learning, deals with multi-dimensional data, which can be efficiently represented using vectors and matrices- the central objects in linear algebra. For instance, an image, which is an input to many AI systems (like those used in facial recognition or autonomous driving), can be represented as a 2D or 3D matrix of pixel intensity values.

Example:

Let's take a simple linear regression problem - a fundamental machine learning task. Here, we try to fit a line to a set of data points in a way that minimizes the distance from each point to the line. The line can be represented as $y = ax + b$, where a is the slope, b is the y-intercept, x is the input, and y is the output. In vectorized form, this equation can be written as $y = Xa$, where X is a matrix of input data and a is a vector of parameters. This problem's solution relies on fundamental operations from linear algebra.

Calculus and AI

Calculus, specifically differential calculus, is integral to AI because it provides the tools for optimization- a key task in machine learning algorithms. At its core, training an AI model involves minimizing a loss function that quantifies the discrepancy between the model's predictions and the actual data. This optimization is usually achieved through an algorithm called gradient descent, which uses derivatives from calculus.

Example:

Consider training a neural network. This process involves iteratively adjusting the network's weights to minimize a loss function. This is done using an algorithm called backpropagation, which is the application of the chain rule from calculus to compute gradients efficiently. Without the concept of derivatives from calculus, training such complex models would not be possible.

Probability Theory and Statistics and AI

Probability theory and statistics form the backbone of many AI algorithms, offering a framework to manage uncertainty and extract knowledge from data. From Naive Bayes classifiers to Markov chains used in reinforcement learning, these concepts play a significant role.

Example: The Bayesian framework, named after Thomas Bayes, forms the basis of many AI algorithms. Bayes' Theorem, a fundamental theorem in probability theory, provides a way to update our beliefs based on new evidence. In a spam filtering algorithm, for example, Bayes' theorem can be used to calculate the probability that an email is spam given the occurrence of certain words in it.

Conclusion- The Interplay of Math and AI

Linear algebra, calculus, and probability theory form the trifecta of mathematical disciplines powering AI. While you can use AI libraries and tools without understanding the underlying math, gaining insight into these mathematical foundations allows you to design better algorithms, troubleshoot issues, and make more informed decisions about which tools and approaches to use.

Math offers the compass that guides you through the vast, often challenging, landscape of AI. By understanding these fundamentals, you can unlock deeper insights into how AI systems work, opening new possibilities for innovation and discovery in this rapidly evolving field.

In the next topics, we will delve deeper into these mathematical concepts and their applications to AI, providing you with the tools you need to navigate the world of AI with confidence and expertise.

Programming Fundamentals for Artificial Intelligence

The field of Artificial Intelligence (AI) has seen significant advancements, propelled by the progress in computer programming. AI algorithms and models are implemented using various programming languages, with Python leading due to its simplicity and the availability of robust scientific computing and machine learning libraries.

To work effectively in AI, one should be well-versed in programming fundamentals such as:

Variables and Data Types: Understanding how to declare and use variables, and the different data types.

Control Structures: Concepts like loops, conditional statements, and exception handling.

Functions: Both built-in and user-defined functions.

Data Structures: Understanding of basic data structures such as arrays, lists, sets, dictionaries, trees, and graphs.

Object-Oriented Programming (OOP): Concepts like classes, objects, inheritance, and polymorphism.

File Input/Output (I/O): Reading from and writing to files.

Variables, Data Types, and Control Structures

Variables and Data Types: In any programming language, variables are used to store information, and each variable is associated with a specific data type. For instance, in Python, we have several data types such as int (integer), float (floating point number), str (string), bool (boolean), etc. Understanding how to work with these data types is crucial in AI programming.

For example, in a machine learning model, features might be represented as floats, labels might be integers (for classification tasks), and file paths (for data loading) might be strings.

```python
# Example
age = 25  # integer
height = 180.5  # float
name = "John Doe"  # string
is_student = False  # Boolean
```

Control Structures: These include conditional statements (if, else, elif in Python) and loops (for, while in Python). Control structures allow your code to make decisions and repeat actions.

```python
# Example
for i in range(10):
    if i % 2 == 0:
        print(f"{i} is even.")
    else:
        print(f"{i} is odd.")
```

Functions and Data Structures

Functions: Functions are reusable pieces of code that perform a specific task. They can be built-in, or user defined. For example, Python's built-in len() function returns the length of a list, and a user-defined function calculate_area() might return the area of a rectangle.

```python
# Example
def calculate_area(width, height):
    return width * height

# Usage
area = calculate_area(5, 10)
print(f"The area of the rectangle is {area}.")
```

Data Structures: These are formats for organizing and storing data. Common data structures include lists, tuples, sets, dictionaries (in Python), as well as more complex ones like trees and graphs. In AI, manipulating data structures is a common task.

```python
# Example
student_grades = {"John": 85, "Emma": 90, "Sophia": 95}  # A
dictionaries in Python
print(student_grades["Emma"])   # Prints: 90
```

Object-Oriented Programming and File I/O

Object-Oriented Programming (OOP): OOP is a programming paradigm that uses "objects" (instances of classes) to design applications. Understanding concepts such as classes, objects, methods, inheritance, and polymorphism is important in AI programming.

```python
# Example
class Rectangle:
    def __init__(self, width, height):
        self.width = width
        self.height = height

    def area(self):
        return self.width * self.height

# Usage
r = Rectangle(5, 10)
print(f"The area of the rectangle is {r.area()}.")
```

File Input/Output (I/O): Reading data from files and writing data to files are common tasks in AI. Data used for training AI models is often stored in files.

```python
# Example
with open('data.txt', 'r') as file:
    data = file.read()
print(data)
```

This code snippet reads the content of 'data.txt' and prints it.

The Importance of Python in AI

Python is often the go-to language in AI because it combines a simple syntax with powerful libraries for data manipulation, machine learning, and scientific computation, like NumPy, pandas, Matplotlib, SciPy, scikit-learn, and TensorFlow.

For instance, here's how easy it is to train a machine learning model using scikit-learn in Python:

```python
from sklearn.ensemble import RandomForestClassifier
from sklearn.datasets import load_iris
from sklearn.model_selection import train_test_split
from sklearn.metrics import accuracy_score

# Load data
iris = load_iris()
X, y = iris.data, iris.target

# Split data into training and test sets
X_train, X_test, y_train, y_test = train_test_split(X, y,
test_size=0.2, random_state=42)

# Create a model
model = RandomForestClassifier()

# Train the model
model.fit(X_train, y_train)

# Make predictions
predictions = model.predict(X_test)

# Measure accuracy
accuracy = accuracy_score(y_test, predictions)
print(f"Model accuracy: {accuracy}")
```

Conclusion- Programming Fundamentals and AI

In conclusion, understanding programming fundamentals is a vital requirement for anyone aiming to work in AI. These concepts form the foundation upon which more complex AI algorithms and models are built. Python, with its rich ecosystem of AI and machine learning libraries, is particularly useful for implementing and experimenting with AI algorithms. However, the programming concepts mentioned here are universal and apply to virtually all programming languages.

In the upcoming chapters, we will dive deeper into the application of these concepts in AI, exploring how to work with data, build AI models, evaluate their performance, and more.

Artificial Intelligence Concepts

Embarking on your AI journey necessitates a deep understanding of a few core AI concepts. These concepts serve as the building blocks of AI systems and applications and are essential to achieving a sound understanding of the AI field. Here, we will discuss:

Machine Learning: The method of creating models that learn from data.

Deep Learning: A subset of machine learning inspired by the structure of the human brain.

Reinforcement Learning: Learning by interacting with an environment.

Natural Language Processing (NLP): AI applied to understanding and generating human language.

Computer Vision: AI applied to understanding and interpreting visual data.

Machine Learning

Machine Learning (ML) is a subset of AI where computers learn from data without being explicitly programmed. ML algorithms find patterns in data and create models, which are used to make predictions or decisions. The process involves providing an ML algorithm with training data to learn from.

Example: The simplest form of a machine learning algorithm is linear regression. It models the relationship between two variables by fitting a linear equation to the observed data. One variable is considered an explanatory variable, and the other is considered a dependent variable.

```python
from sklearn.linear_model import LinearRegression

# Let's suppose we have the following data
X = [[1], [2], [3], [4], [5]]  # explanatory variable
y = [2, 4, 6, 8, 10]  # dependent variable

model = LinearRegression().fit(X, y)
prediction = model.predict([[6]])  # Predict for X=6
print(prediction)  # Output: [12.]
```

The model learns the relationship between X and y during the .fit() step and later uses this learned relationship to predict the output for new input.

Deep Learning

Deep Learning is a subset of ML that involves artificial neural networks with several layers (hence 'deep'). These neural networks attempt to simulate the behavior of the human brain—albeit far from matching its ability— to 'learn' from substantial amounts of data. While a neural network with a single layer can still make approximate predictions, additional hidden layers can help optimize the accuracy.

Example: Let's consider a simple image recognition task where the goal is to classify images of handwritten digits. For this task, we can use a convolutional neural network (CNN), a class of deep learning models particularly effective for image-related tasks.

```python
from keras.datasets import mnist
from keras.models import Sequential
from keras.layers import Dense, Dropout, Flatten
from keras.layers import Conv2D, MaxPooling2D

# Load MNIST dataset
(X_train, y_train), (X_test, y_test) = mnist.load_data()

# Preprocessing steps omitted...

# Define the model architecture
model = Sequential()
model.add(Conv2D(32, kernel_size=(3, 3), activation='relu',
input_shape=input_shape))
model.add(Conv2D(64, (3, 3), activation='relu'))
model.add(MaxPooling2D(pool_size=(2, 2)))
model.add(Dropout(0.25))
model.add(Flatten())
model.add(Dense(128, activation='relu'))
model.add(Dropout(0.5))
model.add(Dense(num_classes, activation='softmax'))

# Compile the model
model.compile(loss=keras.losses.categorical_crossentropy,
optimizer=keras.optimizers.Adadelta(),
metrics=['accuracy'])

# Train the model
model.fit(X_train, y_train, batch_size=batch_size,
epochs=epochs, verbose=1, validation_data=(X_test, y_test))
```

Reinforcement Learning

Reinforcement Learning (RL) is a type of machine learning where an "agent" learns to make decisions by taking actions in an environment to maximize a reward signal. The agent learns from trial and error, receiving rewards or penalties for the actions it performs.

Example: Training a model to play a game like chess is an example of reinforcement learning. The agent (the AI model) makes moves (actions) in the game (environment), with the aim of winning (reward). Over time, the model learns which sequences of moves are more likely to lead to winning the game.

Natural Language Processing

Natural Language Processing (NLP) involves the interaction between computers and human language. It allows computers to understand, interpret, and generate human language in a valuable way.

Example: Sentiment analysis is an application of NLP. It involves determining whether a piece of text (like a product review) is positive, negative, or neutral.

```python
from textblob import TextBlob

text = "I love this product. It works perfectly!"
testimonial = TextBlob(text)
print(testimonial.sentiment)
```

The sentiment property returns a namedtuple of the form Sentiment(polarity, subjectivity). The polarity score is a float within the range [-1.0, 1.0]. The subjectivity is a float within the range [0.0, 1.0] where 0.0 is very objective and 1.0 is very subjective.

Computer Vision

Computer Vision involves teaching computers to "see" and understand the content of digital images or videos. It's a field of study focused on processing, analyzing, and understanding images or video data.

Example:

An example of computer vision in action is the facial recognition feature found in many of today's social media platforms. By learning patterns from a large amount of image data, a computer vision algorithm can identify and recognize faces in new images.

Conclusion- Core AI Concepts

These five core AI concepts - machine learning, deep learning, reinforcement learning, natural language processing, and computer vision- provide the basis for much of the ongoing innovation in AI. They are the engines driving advancements in fields as diverse as healthcare, transportation, entertainment, and commerce, among others.

In the subsequent chapters, we will dive deeper into these concepts, exploring how they work, their applications, and the latest trends. We will also explore practical examples and hands-on exercises to help you gain a solid understanding of these core AI concepts.

Data Structures and Algorithms in AI

Data structures and algorithms form the backbone of any program, including those involving artificial intelligence. Here, we will explore:

Basic Data Structures: Understanding arrays, linked lists, stacks, queues, trees, and graphs.

Searching Algorithms: Binary search and breadth-first and depth-first search in graphs.

Sorting Algorithms: Basic sorting algorithms like quicksort and mergesort.

Graph Algorithms: Specialized algorithms like Dijkstra's and A* for pathfinding in graphs.

Basic Data Structures

Data structures are used to organize and manage data. They dictate how data is stored, organized, and manipulated, and they can affect the efficiency of a program.

Arrays: An array is a contiguous block of memory that holds elements of the same type.

```python
array = [1, 2, 3, 4, 5]  # An array of integers in Python
```

Linked Lists: A linked list is a linear data structure where each element is a separate object. Each "node" contains data and a reference to the next node.

Trees: A tree is a hierarchical data structure composed of nodes, where each node has a value and a list of references to other nodes (its children).

Graphs: A graph is a set of objects (vertices) interconnected by links (edges). They can be used to represent many real-world things such as systems of roads, airline flights from city to city, or even how web pages link to each other.

Searching Algorithms

Searching is the process of finding a particular element in a data structure.

Binary Search: Binary search is an algorithm used to find the position of a specific value within a sorted array.

```python
def binary_search(arr, low, high, x):
    if high >= low:
        mid = (high + low) // 2
        if arr[mid] == x:
            return mid
        elif arr[mid] > x:
            return binary_search(arr, low, mid - 1, x)
        else:
            return binary_search(arr, mid + 1, high, x)
    else:
        return -1
```

Depth-First Search (DFS) and Breadth-First Search (BFS): DFS and BFS are two fundamental graph traversal algorithms. DFS explores as far as possible along each branch before retracing its steps. BFS explores all the vertices at the present "depth" before going on to vertices at the next level.

Sorting Algorithms

Sorting is the process of arranging a list of elements in a particular order.

Quicksort: Quicksort is a divide-and-conquer algorithm that works by selecting a 'pivot' element and partitioning the other elements into two groups, those less than the pivot and those greater than the pivot.

```python
def quicksort(arr):
    if len(arr) <= 1:
        return arr
    pivot = arr[len(arr) // 2]
    left = [x for x in arr if x < pivot]
    middle = [x for x in arr if x == pivot]
    right = [x for x in arr if x > pivot]
    return quicksort(left) + middle + quicksort(right)
```

Mergesort: Mergesort is also a divide-and-conquer algorithm that works by dividing the unsorted list into N sublists, each containing one element (a list of one element is considered sorted), and then repeatedly merges sublists to produce new sorted sublists until there is only one sublist remaining.

Graph Algorithms

Graphs are used to represent networks of communication, data organization, computational devices, the flow of computation, etc.

Dijkstra's Algorithm: This algorithm finds the shortest path from one node to all other nodes in a graph. It's widely used in network routing protocols, most notably IS-IS and Open Shortest Path First (OSPF).

A Algorithm*: This is an informed search algorithm, or a best-first search, meaning that it solves problems by searching among all possible paths to the solution (goal) for the one that incurs the smallest cost (least distance travelled, shortest time, etc.), and among these paths it first considers the ones that appear to lead most quickly to the solution.

Conclusion- Data Structures and Algorithms

Understanding data structures and algorithms is critical to creating efficient AI systems. While the sophistication of AI and machine learning libraries often abstracts away much of this complexity, having a firm grasp of data structures and algorithms is helpful for designing more efficient models, especially when working with large-scale data.

The chosen data structure can have a significant effect on the speed and efficiency of an AI system. Similarly, algorithms play a vital role in processing and managing data, training models, and in the performance and efficiency of AI systems. As you progress deeper into AI, you'll often find that the choice of data structures and algorithms becomes a key part of the solution.

In the next chapters, we'll explore how these concepts are applied in AI technologies, focusing on different areas such as machine learning, deep learning, natural language processing, and computer vision.

Machine Learning Basics

Machine Learning (ML) is a subfield of AI where algorithms learn from and make decisions or predictions based on data. Key topics include:

Types of Machine Learning: Supervised learning, unsupervised learning, and reinforcement learning.

Key Concepts: Training and testing, overfitting and underfitting, bias-variance trade-off.

Fundamental Algorithms: Linear regression, logistic regression, decision trees, and neural networks.

Types of Machine Learning

Supervised Learning: In supervised learning, we have a labeled dataset, and the goal is to learn a mapping function from inputs to outputs. Examples include predicting house prices based on features like location, size, and age (regression) or classifying emails into spam or not spam (classification).

```python
from sklearn.linear_model import LinearRegression

# Suppose we have a dataset `X` with house features and `y`
with corresponding prices
model = LinearRegression()
model.fit(X, y)  # Train the model
```

Unsupervised Learning: In unsupervised learning, we have an unlabeled dataset, and the goal is to find structure in the data. Examples include clustering customers into distinct groups based on their purchasing behavior or reducing the dimensionality of a dataset.

Reinforcement Learning: In reinforcement learning, an agent learns to perform actions in an environment to maximize some reward. For example, a chess-playing AI learns which moves to make (actions) by playing many games (environment) and trying to win (reward).

Key Concepts

Training and Testing: In machine learning, we typically split our dataset into a training set and a test set. The training set is used to train the model, and the test set is used to evaluate its performance.

Overfitting and Underfitting: Overfitting occurs when a model learns the training data too well and performs poorly on unseen data. Underfitting is when a model fails to capture the underlying pattern of the data. Both result in poor predictive performance.

Bias-Variance Trade-off: Bias is the error due to simplistic assumptions in the learning algorithm, leading to underfitting. Variance is the error due to complex models trying to fit the data too closely, leading to overfitting. The bias-variance trade-off is the balance that must be achieved to minimize total error.

Fundamental Algorithms- Linear Regression

Linear Regression is a supervised learning algorithm used for regression tasks. It models the relationship between a scalar dependent variable y and one or more independent variables X.

```python
from sklearn.linear_model import LinearRegression

# Suppose we have a dataset `X` with independent variables
and `y` with the dependent variable
model = LinearRegression()
model.fit(X, y)  # Train the model
predictions = model.predict(X_test)  # Predict values for
the test set
```

Here, the model is trained to find the line (or hyperplane in multiple dimensions) that best fits the data according to a certain criterion, such as minimizing the mean squared error.

Fundamental Algorithms- Logistic Regression

Logistic Regression is a supervised learning algorithm used for binary classification tasks. Despite its name, logistic regression is a classification algorithm that models the probability that a given input belongs to a particular category.

```python
from sklearn.linear_model import LogisticRegression

# Suppose we have a dataset `X` with features and `y` with
binary class labels
model = LogisticRegression()
model.fit(X, y)  # Train the model
predictions = model.predict(X_test)  # Predict classes for
the test set
```

In the code above, the logistic regression model is trained on a binary classification problem and then used to make predictions.

Fundamental Algorithms- Decision Trees

Decision Trees are a type of supervised learning algorithm used for both regression and classification tasks. They work by partitioning the input space into regions, each assigned a prediction value based on the majority class (for classification) or average value (for regression) of the samples in that region.

```python
from sklearn.tree import DecisionTreeClassifier

# Suppose we have a dataset `X` with features and `y` with
class labels
model = DecisionTreeClassifier()
model.fit(X, y)  # Train the model
predictions = model.predict(X_test)  # Predict classes for
the test set
```

In this code, a decision tree is used for a classification problem.

Fundamental Algorithms- Neural Networks

Neural Networks, especially deep learning networks, are a powerful class of models used in machine learning. They are inspired by the biological neural networks in the human brain and are used to model complex patterns and perform tasks that require artificial intelligence.

```python
from keras.models import Sequential
from keras.layers import Dense

# Assuming we have a binary classification problem with 10
input features
model = Sequential()
model.add(Dense(32, input_dim=10, activation='relu'))
model.add(Dense(1, activation='sigmoid'))

model.compile(loss='binary_crossentropy', optimizer='adam',
metrics=['accuracy'])

# Train the model
model.fit(X_train, y_train, epochs=50, batch_size=10)
```

In the code above, a simple neural network is constructed using the Keras library, trained, and then used to make predictions. This type of model can manage more complex patterns in the data compared to the other models discussed previously.

Conclusion- Machine Learning Basics

In conclusion, Machine Learning is a core part of AI, and understanding its basics is essential for any AI practitioner. This includes knowing the several types of machine learning and when to use them, being aware of key concepts like training and testing, overfitting and underfitting, and the bias-variance trade-off, and having a basic understanding of fundamental algorithms.

As we dive deeper into AI in the coming chapters, we will explore more advanced topics in machine learning, including ensemble methods, support vector machines, and deep learning models.

Deep Learning

Deep learning, a subset of machine learning, utilizes neural networks with many layers ("deep" structures) to model and understand complex patterns and representations. In this section, we will explore:

- **Neural Networks and Perceptrons**
- **Convolutional Neural Networks (CNNs)**
- **Recurrent Neural Networks (RNNs)**
- **Generative Adversarial Networks (GANs)**
- **Transfer Learning**

Neural Networks and Perceptrons

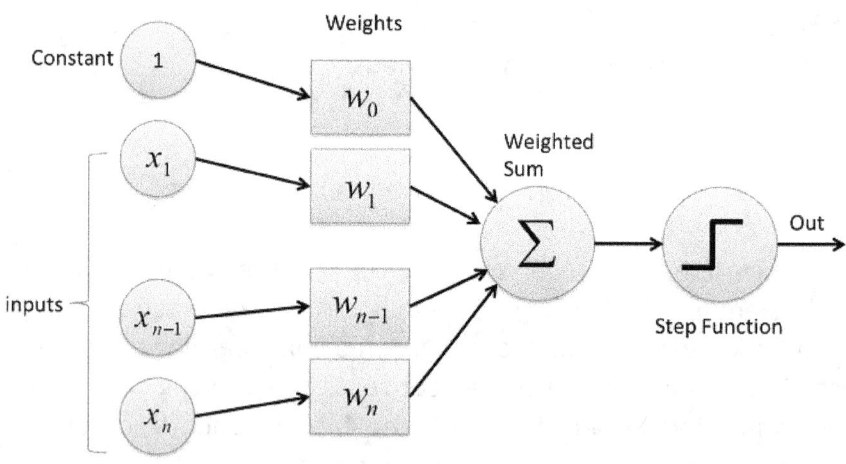

*Source 1. https://miro.medium.com/max/700/1*n6sJ4yZQzwKL9wnF5wnVNg*

A neural network is comprised of layers of nodes or "neurons." Each node in a layer receives inputs, applies a function to these inputs, and passes its output to nodes in the next layer. The simplest kind of neural network is a single-layer perceptron network, which consists of a single layer of output nodes connected to a layer of input nodes.

A single neuron or perceptron can be visualized as follows:

```
from keras.models import Sequential
from keras.layers import Dense

# Define the model
model = Sequential()
model.add(Dense(32, input_dim=10, activation='relu'))  #
Hidden layer with 32 nodes
model.add(Dense(1, activation='sigmoid'))  # Output layer
with 1 node

model.compile(loss='binary_crossentropy', optimizer='adam',
metrics=['accuracy'])
model.fit(X_train, y_train, epochs=50, batch_size=10)  #
Train the model
```

Convolutional Neural Networks (CNNs)

CNNs are a class of deep learning models that are predominantly used in the processing of images. CNNs are designed to automatically and adaptively learn spatial hierarchies of features from the input.

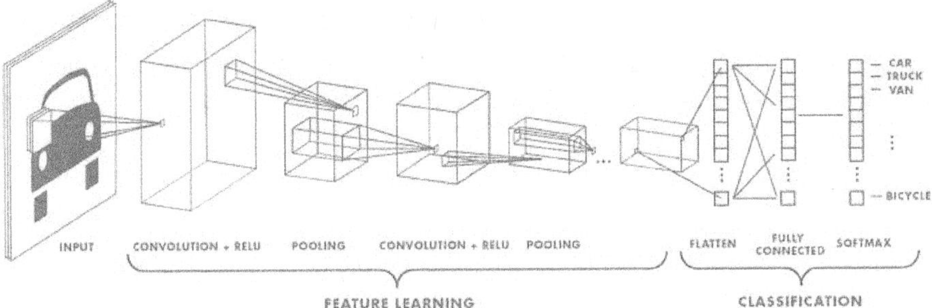

*Source 2. https://miro.medium.com/v2/resize:fit:700/1*vkQ0hXDaQv57sALXAJquxA*

```python
from keras.models import Sequential
from keras.layers import Conv2D, MaxPooling2D, Flatten,
Dense

model = Sequential()
model.add(Conv2D(32, kernel_size=(3, 3), activation='relu',
input_shape=(64, 64, 3)))
model.add(MaxPooling2D(pool_size=(2, 2)))
model.add(Flatten())
model.add(Dense(128, activation='relu'))
model.add(Dense(1, activation='sigmoid'))

model.compile(optimizer='adam', loss='binary_crossentropy',
metrics=['accuracy'])
model.fit(X_train, y_train, epochs=25, batch_size=32)
```

This code defines a simple CNN with one convolutional layer, followed by a max pooling layer, a flattening layer, and two dense layers.

Recurrent Neural Networks (RNNs)

RNNs are a type of neural network designed for sequence prediction problems. They can read inputs of varying length (a series) and return a single output (e.g., sentiment analysis) or a sequence of outputs (e.g., machine translation).

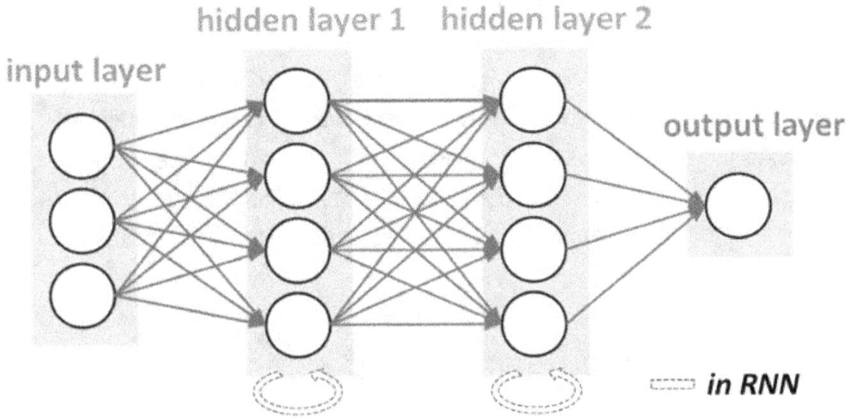

hidden layer 1   hidden layer 2

input layer

output layer

in RNN

Source 2. https://www.researchgate.net/figure/Generalized-recurrent-neural-network-architecture-with-two-hidden-layers-The-NN_fig3_335159004

An example of a simple RNN implemented using Keras:

```python
from keras.models import Sequential
from keras.layers import SimpleRNN, Dense

model = Sequential()
model.add(SimpleRNN(32, input_shape=(time_steps,
n_features), return_sequences=True))
model.add(SimpleRNN(32))
model.add(Dense(1, activation='sigmoid'))

model.compile(loss='binary_crossentropy', optimizer='adam',
metrics=['accuracy'])
model.fit(X_train, y_train, epochs=50, batch_size=32)
```

Here, we have a two-layered RNN that eventually leads to a dense output layer.

Generative Adversarial Networks (GANs)

GANs are a class of AI algorithms used in unsupervised machine learning. They were introduced by Ian Goodfellow and his colleagues in 2014. GANs are used to generate updated content, and they consist of two parts- the Generator, which generates the images, and the Discriminator, which classifies real and fake images.

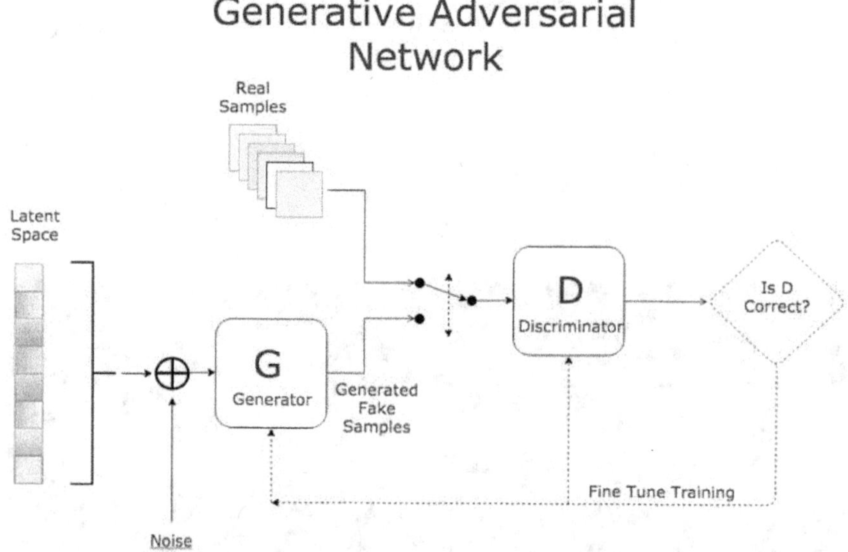

*Source 3. https://miro.medium.com/v2/resize:fit:700/1*5rMmuXmAquGTT-odw-bOpw*

Building and training a GAN is more complex and beyond the scope of a simple example, but a typical GAN training loop involves training the discriminator to distinguish real from fake images, then training the generator to fool the discriminator.

Transfer Learning

Transfer learning is a machine learning method where a pre-trained model is used on a new problem. It is currently exceedingly popular in deep learning because it can train deep neural networks with comparatively little data.

For example, consider a pre-trained CNN like VGG16, trained on ImageNet, used as a feature extractor:

```python
from keras.applications.vgg16 import VGG16
from keras.models import Model

# Load the VGG16 model
base_model = VGG16(weights='imagenet', include_top=False,
input_shape=(200, 200, 3))

# Define a new model with the last layers of VGG16
model = Sequential()
model.add(base_model)
model.add(Flatten())
model.add(Dense(1, activation='sigmoid'))

model.compile(optimizer='adam', loss='binary_crossentropy',
metrics=['accuracy'])
model.fit(X_train, y_train, epochs=10, batch_size=32)
```

Here, we import the VGG16 model, remove its final layers, and add a new output layer that fits our problem's needs.

Conclusion- Deep Learning

Deep learning provides a versatile set of tools that enable us to tackle problems that were previously thought to be too complex to solve. From processing image and text data to generating completely updated content, deep learning models have significantly advanced the field of artificial intelligence.

In the next chapters, we will delve into specific applications of AI, such as natural language processing and

Natural Language Processing (NLP)

Natural Language Processing (NLP) is a subfield of AI that focuses on the interaction between computers and humans using natural language. In other words, NLP is a way for computers to analyze, understand, and derive meaning from human language in a valuable way.

In this section, we will explore the following topics:

- Basics of NLP
- Text Preprocessing Techniques
- Word Embeddings
- NLP Models
- Transformers and Attention Mechanism

Basics of NLP

Fundamental tasks in NLP include:

Tokenization: Breaking text into tokens (words, sentences, etc.)

Part-of-Speech Tagging: Identifying the part of speech of every token.

Named Entity Recognition (NER): Identifying entities (e.g., persons, locations, organizations).

Sentiment Analysis: Determining the sentiment expressed in text.

Topic Modeling: Discovering the main topics in a text.
For instance, in Python, you can use the NLTK library to perform some of these tasks:

```python
import nltk

sentence = "OpenAI is an artificial intelligence research lab."
tokens = nltk.word_tokenize(sentence)  # Tokenization
pos_tags = nltk.pos_tag(tokens)  # Part-of-Speech Tagging

print(f"Tokens: {tokens}")
print(f"POS tags: {pos_tags}")
```

Text Preprocessing Techniques

Text often requires preparation before it can be used in NLP models. Common preprocessing techniques include:

Lowercasing: Converting all the text to lower case.

Stopwords Removal: Removing commonly used words like 'is', 'the', 'and', etc.

Stemming: Reducing a word to its root form.

Lemmatization: Reducing a word to its base form considering the context.

In Python, you can use NLTK for these tasks:

```python
from nltk.corpus import stopwords
from nltk.stem import PorterStemmer, WordNetLemmatizer

# Assuming `tokens` are the tokens from the previous example
tokens = [token.lower() for token in tokens if token not in stopwords.words('english')]  # Lowercasing and stopwords removal
tokens = [PorterStemmer().stem(token) for token in tokens]  # Stemming
tokens = [WordNetLemmatizer().lemmatize(token) for token in tokens]  # Lemmatization
```

Word Embeddings

Word embeddings are a type of word representation that allows words with similar meaning to have a similar representation. They are a distributed representation for text that is perhaps one of the key breakthroughs for the impressive performance of deep learning methods on challenging NLP problems.

Word2Vec and GloVe are two popular models to create such word embeddings.

Here is a code snippet demonstrating how to load pre-trained word2Vec embeddings in Gensim:

```python
from gensim.models import KeyedVectors

# Load vectors directly from the file
model = KeyedVectors.load_word2vec_format('GoogleNews-vectors-negative300.bin', binary=True)

# Access vectors for specific words with a keyed lookup:
vector = model['easy']
# see the shape of the vector (300,)
print(vector.shape)
```

NLP Models

NLP tasks can be solved using various models ranging from traditional machine learning models like Naive Bayes, Support Vector Machine (SVM), to deep learning models like Recurrent Neural Networks (RNNs), Convolutional Neural Networks (CNNs), and more recently transformers-based models like BERT, GPT.

For instance, here is an example of training a Naive Bayes classifier for sentiment analysis in Scikit-Learn:

```python
from sklearn.feature_extraction.text import CountVectorizer
from sklearn.naive_bayes import MultinomialNB
from sklearn.pipeline import Pipeline

# `X_train` is a list of text documents and `y_train` are
the labels
pipeline = Pipeline([
    ('vectorizer', CountVectorizer()),  # Convert the text
data into bag-of-words feature vectors
    ('classifier', MultinomialNB())  # Use Naive Bayes
classifier
])

pipeline.fit(X_train, y_train)
```

Transformers and Attention Mechanism

Transformers are a type of model architecture introduced in the paper "Attention is All You Need" by Vaswani et al. They rely on the attention mechanism, which weighs the input tokens' relevance, providing a boost to the most relevant tokens.

BERT (Bidirectional Encoder Representations from Transformers) and GPT (Generative Pre-trained Transformer) are examples of transformers that have shown state-of-the-art results on many NLP tasks.

Computer Vision

Computer Vision (CV) is a subfield of AI and machine learning concerned with understanding and processing images or videos to produce numerical or symbolic information. It can automate tasks that the human visual system can do.

In this section, we will discuss:

- Fundamentals of Computer Vision
- Image Preprocessing Techniques
- Image Classification
- Object Detection
- Semantic Segmentation

Fundamentals of Computer Vision

Computer Vision tasks include:

Image Classification: Assigning a label to an image from a predefined set of categories.

Object Detection: Identifying the presence, location, and type of one or more objects in an image.

Semantic Segmentation: Assigning a class label to each pixel of the image.

Instance Segmentation: Separating each distinct object of interest from an image.

For instance, the Python library OpenCV provides basic functionalities for many Computer Vision tasks:

```python
import cv2

image = cv2.imread('image.jpg')  # Load an image
gray = cv2.cvtColor(image, cv2.COLOR_BGR2GRAY)  # Convert
the image to grayscale
```

Image Preprocessing Techniques

Image preprocessing is a critical step in Computer Vision. Techniques include:

Grayscale Conversion: Useful for tasks where color does not provide much information.

Normalization: Transforming pixel values to be in a certain range (e.g., 0-1), helping improve the training process.

Resizing: Adjusting images to a certain size.

Data Augmentation: Creating augmented data through transformations like rotations, translations, and flips to improve model performance.

Example of image resizing and normalization using OpenCV:

```python
import cv2
import numpy as np

image = cv2.imread('image.jpg')  # Load an image
image = cv2.resize(image, (64, 64))  # Resize the image
image = image / 255.0  # Normalize the image
```

Image Classification

Image classification involves assigning a class label to an image from a fixed set of categories. The most common approach to an image classification task is the Convolutional Neural Network (CNN).

Here's a simple example of an image classification model using Keras:

```python
from keras.models import Sequential
from keras.layers import Conv2D, MaxPooling2D, Flatten, Dense

model = Sequential()
model.add(Conv2D(32, (3, 3), activation='relu',
input_shape=(64, 64, 3)))
model.add(MaxPooling2D(pool_size=(2, 2)))
model.add(Flatten())
model.add(Dense(128, activation='relu'))
model.add(Dense(10, activation='softmax'))  # Assuming we
have 10 classes

model.compile(optimizer='adam',
loss='categorical_crossentropy', metrics=['accuracy'])
model.fit(X_train, y_train, epochs=10, batch_size=32)
```

Object Detection

Object detection refers to the capability of computer and software systems to locate objects in an image/scene and identify each object. This is more complex than classification, as it also involves localizing the objects within the image. Algorithms include R-CNN, Fast R-CNN, Faster R-CNN, SSD, and YOLO.

The following is an example of using pre-trained YOLO (You Only Look Once) model for object detection:

```python
import cv2

net = cv2.dnn.readNet('yolov3.weights', 'yolov3.cfg')  #
Load the YOLO network
layer_names = net.getLayerNames()
output_layers = [layer_names[i[0] - 1] for i in
net.getUnconnectedOutLayers()]

# Assuming `image` is the image we want to process
blob = cv2.dnn.blobFromImage(image, 0.00392, (416, 416),
(0, 0, 0), True, crop=False)
net.setInput(blob)
outs = net.forward(output_layers)  # Perform forward pass
```

Here 'outs' will contain the information about the detected objects, their position, and the confidence in the detection.

Semantic Segmentation

Semantic segmentation involves labeling each pixel in the image with a category label. Unlike object detection, which provides a bounding box around the object, semantic segmentation provides a precise delineation of the object within the image. Algorithms used for semantic segmentation include FCN, U-Net, and Mask R-CNN.

Here is an example of using pre-trained U-Net for semantic segmentation:

```python
from keras.models import load_model
from keras.preprocessing.image import img_to_array,
load_img

model = load_model('unet_model.h5')  # Load the pre-trained
U-Net model

# Assuming `image_path` is the path to the image we want to
process
image = load_img(image_path, target_size=(256, 256))  #
Load the image and resize it to the expected model input
size
image = img_to_array(image)  # Convert the image to NumPy
array
image = image / 255.0  # Normalize the image
image = np.expand_dims(image, axis=0)  # Add an extra
dimension because the model expects batches of images

pred = model.predict(image)  # Perform prediction
```

In this case, 'pred' will be a 256x256 image where the value of each pixel corresponds to the predicted class of that pixel.

Conclusion- Computer Vision

From classifying images to detecting and precisely segmenting objects within those images, computer vision algorithms have a wide range of applications, including autonomous vehicles, medical imaging analysis, surveillance

Reinforcement Learning

Reinforcement Learning (RL) is a type of machine learning where an agent learns to make decisions by taking actions in an environment to achieve a goal. The agent receives rewards or penalties for the actions it performs, striving to maximize the total reward over time.

In this section, we will delve into:

- Reinforcement Learning Basics
- Markov Decision Processes
- Q-Learning
- Deep Reinforcement Learning
- Applications of Reinforcement Learning

Reinforcement Learning Basics

The main components of a reinforcement learning system are:

Agent: The learner or decision-maker.

Environment: Everything the agent interacts with.

Action (A): What the agent can do.

State (S): The current situation returned by the environment.

Reward (R): Immediate return sent back from the environment to evaluate the last action.

The agent learns a policy π, which is a mapping from states to actions ($\pi: S \to A$), aiming to maximize the cumulative reward over time, known as the return G_t.

Markov Decision Processes

Reinforcement learning problems are often modeled as Markov Decision Processes (MDPs). An MDP is a tuple (S, A, P, R, γ) where:

S is a finite set of states,
A is a finite set of actions,
P is a state transition probability matrix,
R is a reward function, and
γ is the discount factor $\gamma \in [0, 1]$.

The goal in an MDP is to find an optimal policy π^* that maximizes the expected cumulative discounted reward over time.

Q-Learning

Q-Learning is a popular model-free reinforcement learning algorithm. It directly learns a policy that tries to maximize the total reward based on the Q-values, which represent the "quality" of an action taken in a certain state.

Here is a simple example of Q-Learning using the OpenAI Gym environment:

```python
import gym
import numpy as np

env = gym.make('FrozenLake-v0')  # Create environment
Q = np.zeros([env.observation_space.n, env.action_space.n])
# Initialize Q-table with zeros

# Hyperparameters
alpha = 0.5
gamma = 0.95
num_episodes = 5000

# Q-Learning process
for i in range(num_episodes):
    state = env.reset()  # Reset the environment
    done = False

    while not done:
        action = np.argmax(Q[state, :] + np.random.randn(1,
env.action_space.n)*(1./(i+1)))  # Choose an action
        new_state, reward, done, _ = env.step(action)  #
Get new state and reward from environment

        # Update Q-table
        Q[state, action] = Q[state, action] + alpha *
(reward + gamma * np.max(Q[new_state, :]) - Q[state,
action])
        state = new_state  # Update the state
```

Deep Reinforcement Learning

Deep Reinforcement Learning combines reinforcement learning with deep learning. It uses neural networks to approximate the Q-value function in Q-Learning, which is necessary when dealing with larger environments.

One popular Deep RL algorithm is Deep Q-Networks (DQN) introduced by Google's DeepMind. They used a Convolutional Neural Network (CNN) to play Atari games directly from pixel inputs.

Applications of Reinforcement Learning

Reinforcement Learning has many practical applications, including:

Game Playing: Reinforcement learning can be used to train agents that play video games. AlphaGo, developed by DeepMind, used RL to defeat world champion Go players.
Robotics: RL can be used to train robots to perform tasks like walking, running, or complex manipulations.
Resource Management: In systems where resources must be managed over time, like data center cooling, RL can optimize decision-making processes.

Conclusion- Reinforcement Learning

Reinforcement Learning, with its rich theory and growing number of successful applications, is a critical component of modern AI. It poses the problem of learning from interaction to achieve a goal, mapping situations to actions to maximize a reward signal. As we advance in RL research and implementation, the potential impact on various aspects of society is vast, from more efficient resource allocation to advanced autonomous systems.

Probabilistic Reasoning and Bayesian Networks

Probabilistic reasoning is a cornerstone of artificial intelligence, used to deal with uncertainty in the world. Bayesian Networks, also known as Belief Networks or Directed Acyclic Graphical Models, are a powerful tool used for this purpose.

In this section, we'll cover:

- Basics of Probabilistic Reasoning
- Introduction to Bayesian Networks
- Inference in Bayesian Networks
- Learning Bayesian Networks
- Applications of Bayesian Networks

Basics of Probabilistic Reasoning

Probabilistic reasoning allows us to reason and make decisions under uncertainty. It's based on the laws of probability theory, which provides a solid foundation for dealing with uncertain knowledge.

Key concepts in probabilistic reasoning include:

Probability: A measure of the likelihood that an event will occur.

Conditional Probability: The probability of an event given that another event has occurred.

Bayes' Rule: A formula that describes how to update the probabilities of hypotheses when given evidence.

Introduction to Bayesian Networks

A Bayesian Network (BN) is a graphical model that represents the probabilistic relationships among a set of variables. It consists of:

A Directed Acyclic Graph (DAG) where nodes represent variables, and edges represent direct dependencies.
A set of Conditional Probability Distributions (CPDs), one for each variable given its parents in the DAG.
For example, consider a simple Bayesian Network for diagnosing a disease given symptoms. Nodes may represent the presence or absence of certain symptoms and diseases. Edges could represent causal relationships (e.g., a disease-causing certain symptoms).

Inference in Bayesian Networks

Inference in Bayesian Networks involves computing the posterior probabilities of certain variables given evidence. Exact inference can be computationally expensive but can be done using algorithms such as Variable Elimination or the Junction Tree algorithm.

Approximate inference techniques, such as Monte Carlo sampling methods, are often used when exact computation is infeasible.

An example of Bayesian inference is predicting the likelihood of having a disease (D) given a symptom (S):

```
P(D|S) = P(S|D) * P(D) / P(S)
```

Here, P(D|S) is the posterior probability we want to compute, P(S|D) is the likelihood, P(D) is the prior probability of the disease, and P(S) is the marginal likelihood of the symptom.

Learning Bayesian Networks

The structure and parameters of a Bayesian Network can be learned from data. The structure learning can be challenging due to the exponential search space of possible DAGs. Algorithms for structure learning include the Chow-Liu algorithm for tree-structured BNs and various search-and-score methods for general BNs.

Parameter learning, i.e., estimating the CPDs, can be done using Maximum Likelihood Estimation (MLE) if the structure is known and the data is fully observed.

Applications of Bayesian Networks

Bayesian Networks are used in various domains, including:

Medical Diagnosis: BNs can model the probabilistic relationships between diseases and symptoms to aid diagnosis.

Risk Management and Decision Analysis: BNs can help understand and mitigate risks in various fields like finance, project management, and cybersecurity.

Natural Language Processing: In computational linguistics, BNs can model the semantic, syntactic, and discourse dependencies between words and phrases.

Conclusion- Probabilistic Reasoning and Bayesian Networks

Probabilistic reasoning and Bayesian Networks play a vital role in dealing with uncertainty, learning from data, and making informed decisions under uncertainty in AI systems. The ability to model complex dependencies, perform sound probabilistic reasoning, and learn from data makes Bayesian Networks a powerful tool for many AI applications.

Search Algorithms

Search algorithms form the foundation of many artificial intelligence applications, enabling systems to explore a solution space to find the best path or the best state according to defined criteria.

This section will cover:

- Introduction to Search Algorithms
- Uninformed Search Strategies
- Informed Search Strategies
- Adversarial Search and Game Theory
- Applications of Search Algorithms

Introduction to Search Algorithms

Search algorithms operate on a state space, which is a formal model of a problem that includes a set of states and a set of actions that transition between these states. The objective of a search algorithm is to find a sequence of actions (a path) from an initial state to a goal state.

Key terms include:

State: A representation of a physical configuration.
Action: A description of the process that can change one state to another.
Path: A sequence of states connected by a sequence of actions.

Uninformed Search Strategies

Uninformed (or blind) search strategies do not have additional information about states or the goal beyond the problem definition. Key methods include:

Breadth-First Search (BFS): BFS explores all the nodes at the present depth before moving on to nodes at the next depth level.

Depth-First Search (DFS): DFS explores as far as possible along each branch before backtracking.

Uniform Cost Search (UCS): UCS is a type of BFS where the least "costly" node is expanded first.

Informed Search Strategies

Informed search strategies use problem-specific knowledge to find solutions more efficiently. Key methods include:

Greedy Best-First Search: This strategy selects the most promising node based on a heuristic function h(n), which estimates the cost from node n to the goal.

A Search*: A* is an extension of UCS that uses a heuristic function to estimate the cost to the goal from the node n. A* is optimal and complete given certain conditions on h(n).

Adversarial Search and Game Theory

Adversarial search is a significant area of AI that studies how decision-making algorithms can be designed for situations where multiple agents are in conflict, such as in games. Game theory provides the theoretical foundation for such problems.

Minimax Algorithm: The Minimax algorithm is a recursive algorithm used for decision-making in two-player games. It assumes optimal play from both players and explores the game tree to minimize the worst-case scenario (hence the name "minimax").

Alpha-Beta Pruning: This is an optimization technique for the minimax algorithm that eliminates branches in the game tree that don't need to be searched because there's already a better move available.

Applications of Search Algorithms

Search algorithms are used extensively in various AI applications:

Pathfinding: In robotics and video games, search algorithms are used to find the shortest path between two points.

Solving Puzzles: Search algorithms can be used to solve puzzles like the 8-puzzle or Rubik's Cube.

Planning: In AI, search algorithms can be used to sequence actions to achieve a goal.

Conclusion- Search Algorithms

From finding routes on a map, solving puzzles, to making strategic decisions in complex games, search algorithms play a crucial role in various AI systems. By learning about these methods, one gains important skills in formulating problems, understanding the complexities, and implementing efficient solutions.

Knowledge Representation and Reasoning

Knowledge Representation and Reasoning (KR&R) is a fundamental aspect of artificial intelligence that focuses on how to symbolize knowledge in a way that a computer system can use to solve complex tasks.

In this section, we will cover:

- Basics of Knowledge Representation
- Logic-Based Knowledge Representation
- Semantic Networks and Frames
- Ontologies
- Applications of Knowledge Representation

Basics of Knowledge Representation

Knowledge Representation is about creating computer-friendly structures that represent data or knowledge. The aim is to design symbolic structures that:

Represent: Encode pieces of information.

Interpret: Can be used by AI systems to understand and reason.

Process: Allow systems to manipulate the knowledge to solve problems.

Logic-Based Knowledge Representation

Logic-based Knowledge Representation uses formal logic to represent knowledge and perform reasoning. It provides a well-defined syntax and semantics, allowing us to formulate statements that are either true or false.

Propositional Logic: Each statement, or "proposition", is either true or false, but not both.

First-Order Logic: This extends propositional logic by introducing variables, quantifiers, and predicates.

For example, in first-order logic, we can represent the statement "All humans are mortal" as "For all x, if x is a Human, then x is Mortal."

Semantic Networks and Frames

Semantic Networks and Frames are knowledge representation techniques that use a graph-like or object-oriented approach.

Semantic Networks: Represent knowledge as a graph where nodes are concepts and edges represent relationships between concepts.

Frames: Represent knowledge as collections of attributes and values to describe entities, like an object-oriented approach.

For example, in a semantic network, we could have "Bird" as a node, and "Can Fly" as an edge connecting to another node "Animal".

Ontologies

An ontology is a formal, explicit specification of a shared conceptualization. It provides a shared and mutual understanding of a domain that can be communicated between people and AI systems. Ontologies include the representation of categories, properties, and relations between concepts.

For example, the ontology for a university might include concepts such as "Student", "Professor", and "Course", and the relationships between them such as "enrolled in", "teaches", etc.

Applications of Knowledge Representation

Knowledge Representation and Reasoning techniques are used in a wide range of applications:

Expert Systems: AI systems that apply reasoning capabilities to reach conclusions.

Natural Language Processing: To understand and generate natural language.

Semantic Web: To create a web of data that can be processed by machines.

Conclusion- Knowledge Representation and Reasoning

The ability to effectively represent knowledge is crucial to creating intelligent systems that understand, learn, and respond to complex scenarios. From Logic to Ontologies, the field of Knowledge Representation and Reasoning offers numerous tools and techniques to structure and use knowledge in a meaningful way.

AI in Practice

Artificial Intelligence is not just a theoretical field; it has numerous real-world applications transforming industries and everyday life.

This section will cover:

- AI in Business
- AI in Healthcare
- AI in Autonomous Vehicles
- AI in Entertainment
- Ethical Considerations in AI

AI in Business

AI is playing an increasingly significant role in business:

Automating Business Processes: AI can automate repetitive tasks, improving efficiency and accuracy. For example, Robotic Process Automation (RPA) uses AI to automate data entry tasks.

Customer Experience: Chatbots and virtual assistants use natural language processing and machine learning to provide personalized customer service.

Predictive Analytics: Machine learning models can analyze historical data to forecast future trends, helping businesses make informed decisions.

AI in Healthcare

AI has immense potential in healthcare:

Medical Imaging: AI algorithms can analyze medical images, like X-rays and MRIs, to detect diseases such as cancer at initial stages.

Personalized Medicine: AI can analyze a patient's genetic information to recommend personalized treatment plans.

Drug Discovery: AI can help identify potential new drugs and streamline clinical trials, reducing the time and cost of drug development.

AI in Autonomous Vehicles

AI is the backbone of autonomous vehicles:

Perception: AI algorithms interpret sensor data to identify objects, roads, and traffic signs.

Path Planning: AI is used to determine the safest and most efficient route to the destination.

Control: AI systems continuously monitor and adjust the vehicle's movements to ensure a safe and smooth ride.

AI in Entertainment

The entertainment industry is harnessing the power of AI:

Game Development: AI is used to create intelligent characters, procedural content, and to personalize player experiences.

Film Production: AI can automate certain aspects of film production, such as editing and sound design.

Music: AI algorithms can compose music and assist in the creation of new compositions.

Ethical Considerations in AI

The growing impact of AI has brought various ethical issues to the forefront:

Privacy: AI systems often require copious amounts of data, raising concerns about data privacy and security.

Bias: AI models can unintentionally perpetuate and amplify societal biases present in their training data.

Job Displacement: The automation of tasks by AI could lead to job losses in certain sectors.

Conclusion- AI in Practice

From business to healthcare, autonomous vehicles to entertainment, AI is transforming our world. But with its capabilities come responsibilities and challenges. Balancing the benefits and addressing the ethical implications of AI will shape how we harness its power in the years to come. As we move forward, it is essential to consider not just what AI can do, but also what it should do.

Ethics and AI

As AI systems become increasingly integrated into our lives, the ethical implications of their use become more pressing. This section will explore the ethical dimensions of AI, including:

- Bias and Fairness
- Transparency and Explainability
- Privacy and Security
- Accountability
- AI and Employment

Bias and Fairness

AI systems can reflect and even amplify societal biases present in their training data. This can lead to discriminatory outcomes. For instance, an AI system used for hiring might favor certain demographics if historical hiring data exhibits this bias.

Mitigating AI bias involves:

Diverse Data Collection: Ensuring the training data represents all groups fairly.

Algorithmic Fairness Techniques: Applying methods to modify algorithms to reduce bias.

Transparency and Explainability

The workings of complex AI models can often be inscrutable, leading to them being dubbed "black boxes." It's crucial for stakeholders to understand how decisions are being made.

Explainable AI (XAI): Techniques designed to make AI decision-making understandable to humans.

Transparency: Disclosing sufficient information about how an AI system operates.

Privacy and Security

AI systems often require enormous amounts of personal data, raising significant privacy concerns.

Data Anonymization: Techniques to remove personally identifiable information.

Differential Privacy: Adding statistical noise to query results to protect individual data privacy.

AI systems are also targets for adversarial attacks, highlighting the need for robust security measures.

Accountability

Who is responsible when an AI system makes a mistake or causes harm?

Regulation: Establishing laws that set standards and repercussions for harmful AI behavior.

Human oversight: Ensuring human involvement in AI decision-making, particularly in critical areas like healthcare or criminal justice.

AI and Employment

AI automation could displace jobs, particularly in sectors like manufacturing or transportation.

Reskilling: Training programs for workers to move into new roles.

Universal Basic Income: A proposed solution for potential widespread job displacement.

Conclusion- Ethics and AI

The ethical considerations of AI are vast and complex. As AI continues to develop, ongoing dialogue and collaboration between technologists, policymakers, and society will be critical to navigate these ethical challenges and ensure AI is used for the benefit of all.

Future of AI

As we continue to advance in our understanding and capabilities, the future of AI holds exciting possibilities. In this section, we will discuss:

- Advances in AI Techniques
- AI and Quantum Computing
- AI in Space Exploration
- Superintelligence
- The Role of AI in Society

Advances in AI Techniques

AI research is in a state of continual evolution, with new techniques and improvements emerging regularly.

Hybrid Models: Combining different AI approaches, such as symbolic reasoning with neural networks, could lead to more robust AI systems.

Neurosymbolic AI: This approach aims to merge deep learning with symbolic reasoning to create AI systems that can reason and learn from less data.

AI and Quantum Computing

Quantum computing is a technology that could significantly accelerate certain computations, and its combination with AI might lead to breakthroughs.

Quantum Machine Learning: Utilizing quantum computers to speed up training and execution of machine learning algorithms.

Quantum Optimization: Solving complex optimization problems, like those found in logistics or finance, more efficiently.

AI in Space Exploration

AI has the potential to revolutionize space exploration.

Autonomous Navigation: AI could guide spacecraft to autonomously navigate and adapt to unexpected situations in space.

Astrobiology: AI could analyze astronomical and planetary data to identify potential signs of extraterrestrial life.

Superintelligence

The concept of superintelligence refers to a form of AI that surpasses human intelligence in virtually all economically valuable work.

AI Safety Research: Before reaching such a stage, considerable research needs to be done to ensure that superintelligent AI aligns with human values and interests.

Long-term Impact: The societal effects of superintelligence are uncertain and warrant careful consideration.

The Role of AI in Society

As AI becomes more advanced and prevalent, its impact on society will grow.

AI Governance: Ensuring that AI is used ethically and responsibly will be a significant challenge.

AI Education: As AI becomes integral to many facets of life, understanding AI will become increasingly important for everyone.

Conclusion- Future of AI

The future of AI is both promising and challenging. It holds the potential for great advancements and improvements in many areas of life, but also poses substantial ethical and societal questions. As we look to the future, it's crucial to navigate these developments with a balanced view of the benefits and potential risks. Our aim should be to use AI to augment human capabilities and to create a future where AI serves the common good.

A Special Note to Our Dear Readers ...

Thank you for embarking on this enlightening journey with us through "Essential Artificial Intelligence for Rookies." We hope that you've found this voyage into the heart of AI not just educational, but truly transformative. But the exploration doesn't end here. Just as AI continually evolves and expands, so too should our understanding of this groundbreaking field.

We are thrilled to announce the upcoming sequel to this book, "Advanced Artificial Intelligence: A Deep Dive into the Digital Mindscape." This next volume is meticulously designed to provide you with an in-depth understanding of the topics we've introduced here. We'll delve deeper into the core mechanics, mathematics, and methodologies that power AI. From dissecting intricate neural networks to decoding complex algorithms, the next book promises to unravel the pure details of AI in an engaging and accessible manner.

As we continue to demystify the future, we encourage you to stay curious, keep asking questions, and persist in your quest for knowledge. This is just the beginning of your AI journey, and the horizon is boundless. Join us as we continue to explore and illuminate the mesmerizing world of artificial intelligence.

Thank you very much.

Mikhail Van Starren